100

Words About

ANIMALS

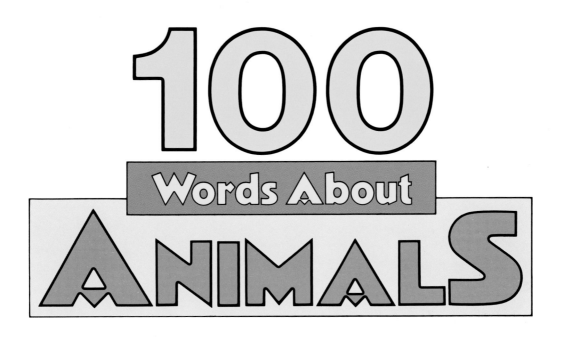

100 Words About ANIMALS

Illustrated by Richard Brown

A Voyager/HBJ Book

Harcourt Brace Jovanovich, Publishers

San Diego   New York   London

Requests for permission to make copies of any
part of the work should be mailed to:
Permissions, Harcourt Brace Jovanovich, Publishers,
Orlando, Florida 32887.

Library of Congress Cataloging-in-Publication Data
100 words about animals.
"Gulliver books."
Summary: Labeled illustrations depict animals in
their natural surroundings, grouped according to
habitats such as farm, sea, and jungle.
1. Animals—Terminology—Juvenile literature.
2. Animals—Pictorial works—Juvenile literature.
3. Picture dictionaries, English—Juvenile literature.
[1. Animals—Pictorial works. 2. Vocabulary]
I. Brown, Richard Eric, 1946–   ill.
II. Title: One hundred words about animals.
QL49.B74 1987   591   86-22744
ISBN 0-15-200550-1
ISBN 0-15-200554-4 (pbk.)
Designed by G.B.D. Smith
Printed and bound by Tien Wah (PTE.) Ltd. Lithographers, Singapore.
A B C D E
A B C D E (pbk.)

To Lynne
 —R.B.

parakeet

guinea pig

hamster

ON THE FARM

cow

sheep

horse

goat

pig

donkey

hen

rooster

turkey

IN THE FIELDS

hawk

skunk

rabbit

mouse

mole

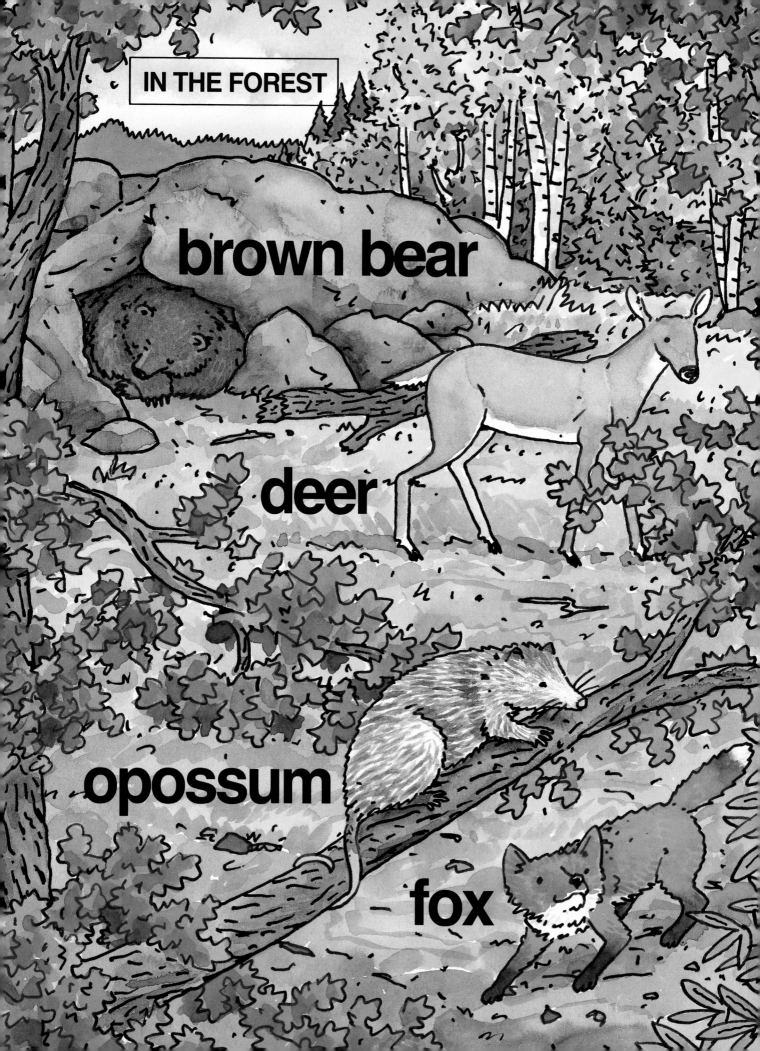

IN THE FOREST

brown bear

deer

opossum

fox

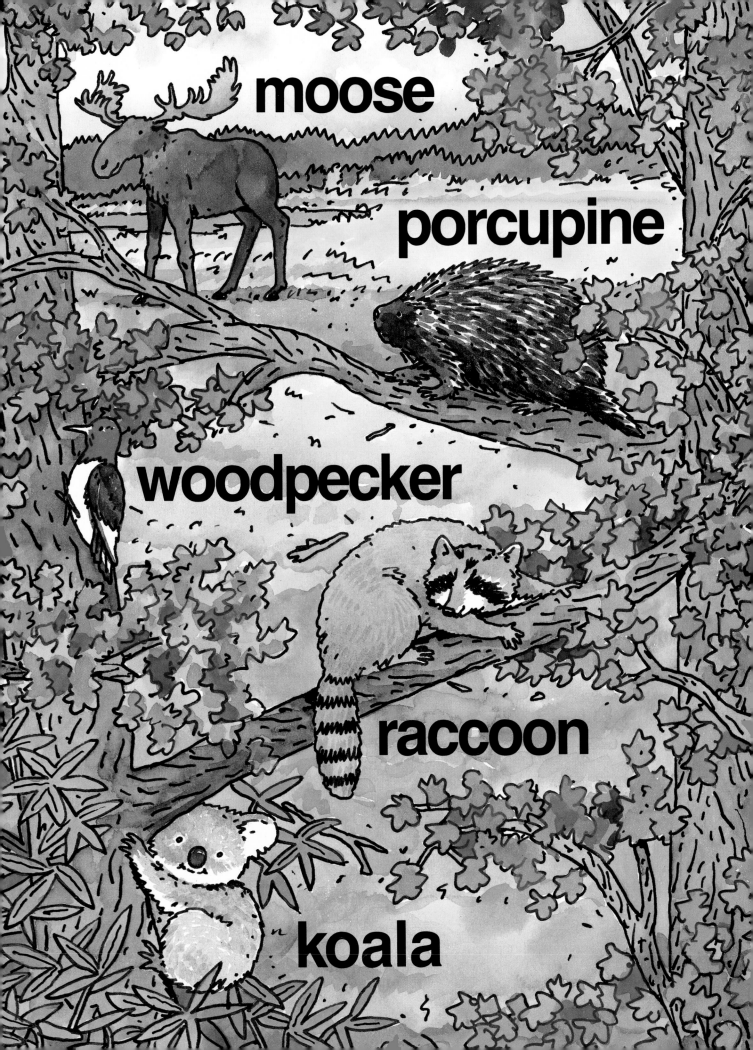

camel

coyote

lizard

whale

shark

seal

sea gull

pelican

sand crab

starfish

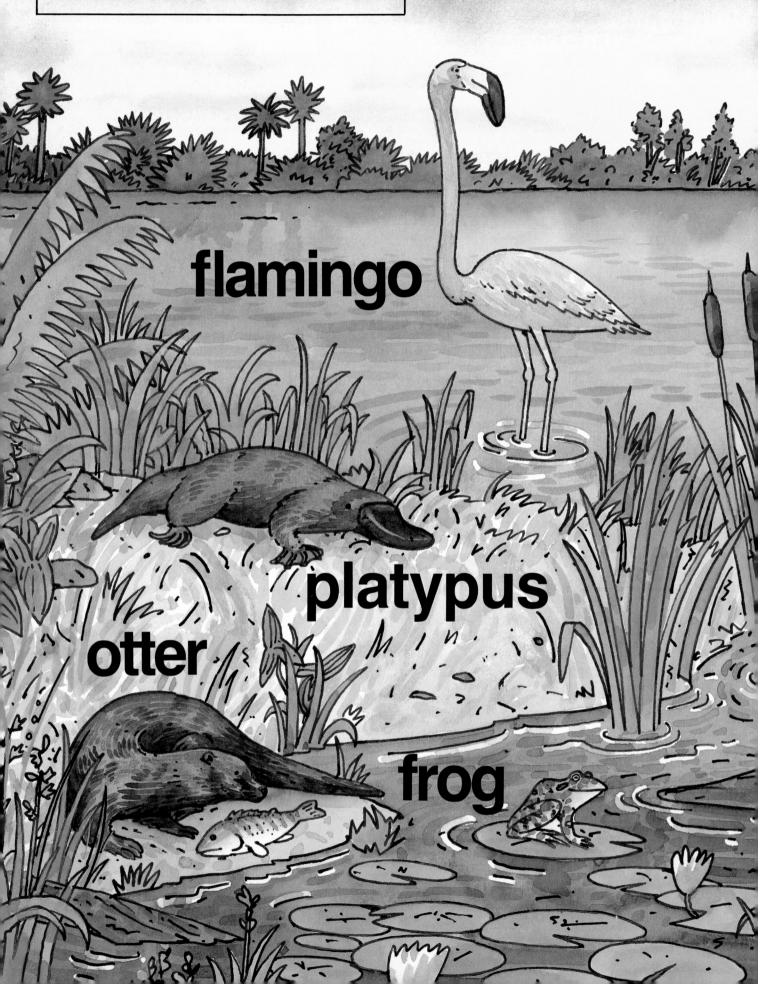

flamingo

platypus

otter

frog

IN THE MOUNTAINS

elk

eagle

timber wolf

yak

llama

panda

musk-ox

reindeer

ermine

hare

polar bear

walrus

penguin

IN THE JUNGLE

parrot

toucan

tiger

crocodile

leopard

gorilla

monkey

anteater

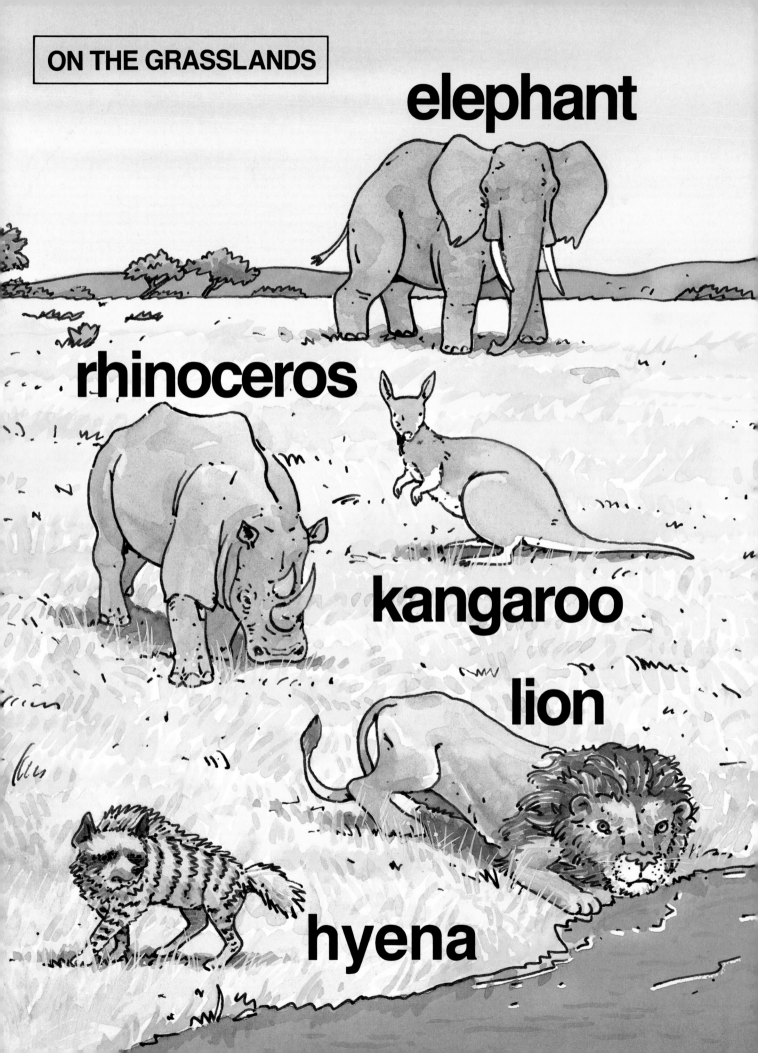

elephant

rhinoceros

kangaroo

lion

hyena

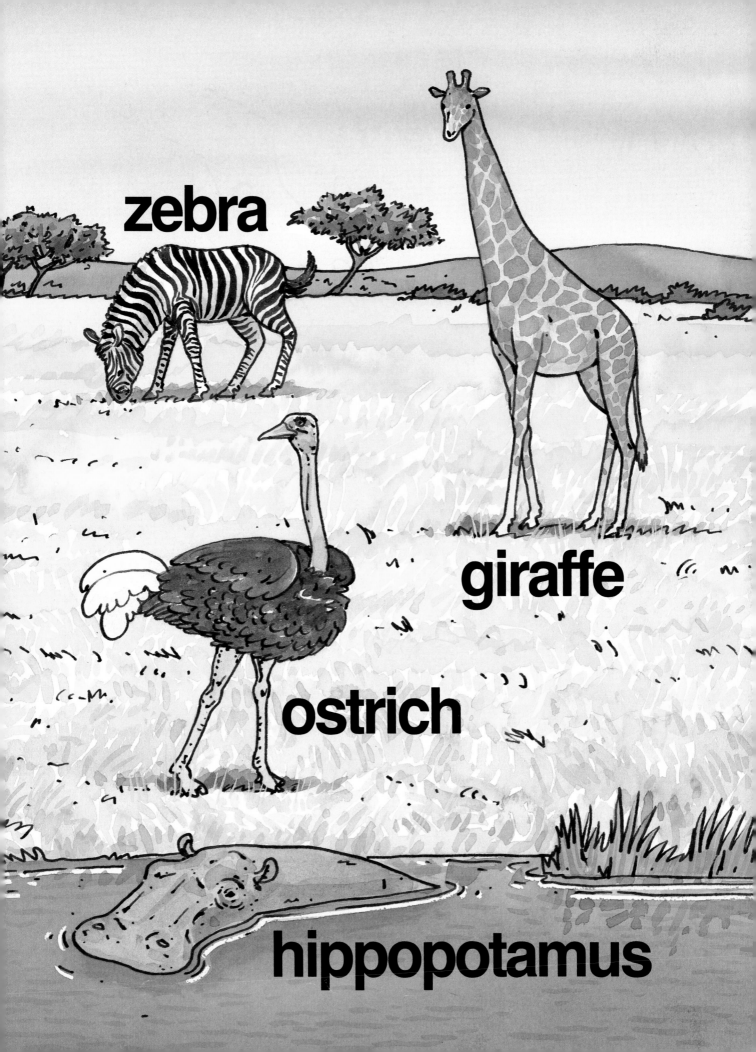

zebra

giraffe

ostrich

hippopotamus

pterodactyl

stegosaurus

tyrannosaurus rex

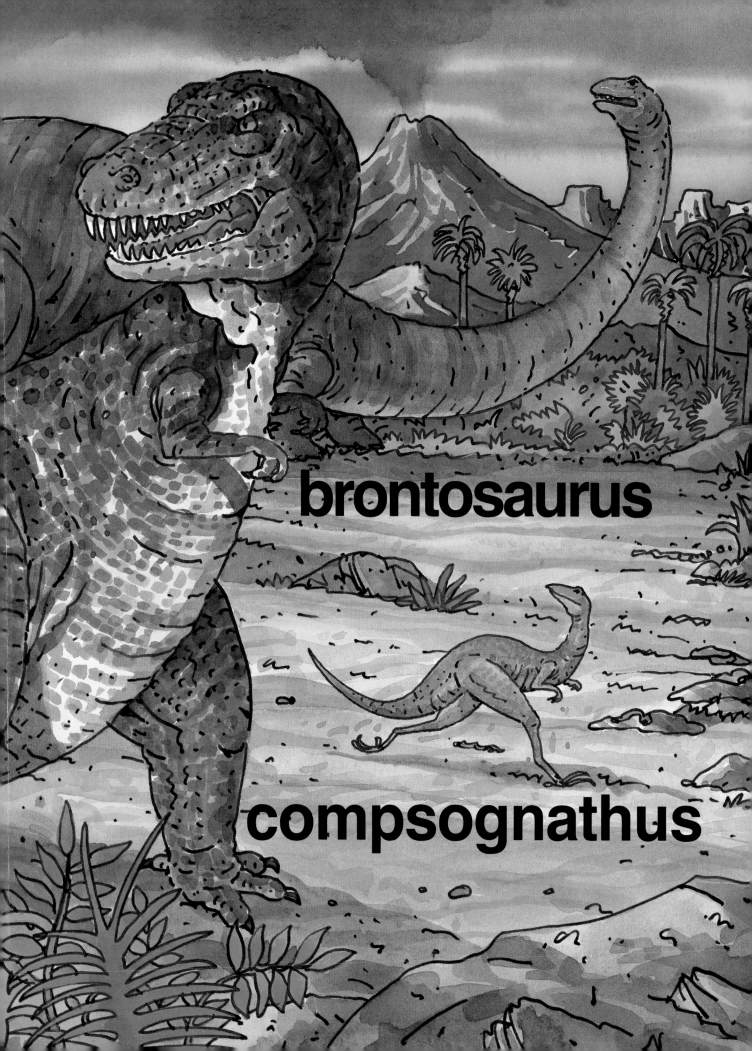

brontosaurus

compsognathus